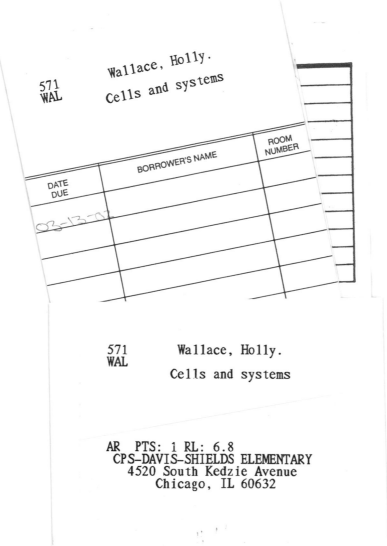

571
WAL

Wallace, Holly.

Cells and systems

DATE DUE	BORROWER'S NAME	ROOM NUMBER
03-13-0?		

571
WAL

Wallace, Holly.

Cells and systems

AR PTS: 1 RL: 6.8

Cells and Systems

Holly Wallace

Heinemann Library
Chicago, Illinois

Designed by Celia Floyd
Originated by Dot Gradations
Printed by Wing King Tong, in Hong Kong

05 04 03 02 01

10 9 8 7 6 5 4 3 2 1

Library of Congress Cataloging-in-Publication Data

Wallace, Holly, 1961-
  Cells and systems / Holly Wallace.
     p. cm. -- (Life processes)
  Includes bibliographical references and index.
  ISBN 1-57572-336-0 (library)
    1. Physiology--Juvenile literature. 2. Cells--Juvenile literature. 3. Tissues--Juvenile literature. 4. Organs (Anatomy)--Juvenile literature. [1. Cells. 2. Organs (Anatomy) 3. Physiology.] I. Title. II. Series.

  QP37 .G333 2000
  571--dc21

                                                          00-040972

Acknowledgments
The author and publishers are grateful to the following for permission to reproduce copyright material: Action Plus/Glyn Kirk, p.15; Neil Bromhall, p. 28; Dr. Jeremy Burgess, p.10; CNRI, p. 7; Dept. of Anatomy, University La Sapienza, Rome/Professor P. Motta, pp.13, 25; Dept. of Anatomy, University La Sapienza, Rome/Professors P. Motta, P.M. Andrews, K.R. Porter, & J. Vial, p. 14; Eye of Science, p. 6; East Anglian Genetics Service/L. Willatt, p. 29; Don Fawcett, p. 27; Manfred Kage, pp. 12, 19; National Cancer Institute, p. 16; Dr. Yorgos Nikas, p. 28; Claude Nuridsany & Marie Perennou, p. 6; D. Phillips, p. 21; Quest, p. 23; Science Photo Library, p. 20; Science Photo Library/Andrew Syred, pp. 5, 11; Don Wong, p. 24.

Cover photograph reproduced with the permission of Science Photo Library.

Every effort has been made to contact copyright holders of any material reproduced in this book. Any omissions will be rectified in subsequent printings if notice is given to the publisher.

Some words are shown in bold, **like this.** You can find out what they mean by looking in the glossary.

Contents

Introduction

Cells and Systems looks at the world of plant and animal cells and systems. Every living thing is made up of cells that form the different parts of its body. In many living things, groups of specialized cells make tissues, such as skin or bones. Groups of tissues make organs, like the heart or lungs. Organs work together to form body systems, which carry out the processes and functions of a living thing.

What Are Cells?

All living things are made up of cells. A cell is a tiny unit of living material that is the basic building block of life. Cells are like tiny factories where chemical reactions happen. These reactions keep living things alive. Plant and animal cells do similar jobs, such as taking in food, releasing energy, and getting rid of waste, but there are some differences in their structures.

Did you know?
Your body is made up of more than 10 trillion cells. Most cells can only be seen under a microscope, but some plant cells can be seen by the human eye. Cells do not grow. When you grow taller it is because your cells have divided so your body has more cells.

Animal cells

An animal cell is like a tiny, **gelatin**-filled bag, held together by a thin, outer cover called a **membrane.** The **nucleus,** which controls everything that happens inside the cell, is found in the center. This diagram shows the main parts of a typical animal cell.

The cell membrane is the cell's outer cover that holds the cell together. It allows food and **oxygen** into the cell, and pushes waste products made in the cell out of it.

Cytoplasm is a gelatin-like substance, made of **protein** and water, that fills most of the cell. The chemical reactions that keep the cell alive happen here.

The nucleus is the control center of the cell. It contains chemical instructions that tell the cell what to do. The nucleus also divides to make new cells for growth, repair, and **reproduction.**

4

Plant cells

Plant cells have all the same features as animal cells with three extra ones. These are a tough, rigid cell wall made of a substance called **cellulose, chloroplasts** for making food, and a large, permanent **vacuole** filled with cell **sap.** Here you can see the main parts of a typical plant cell.

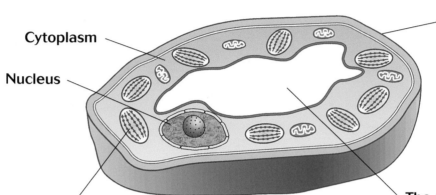

Cytoplasm

Nucleus

The cell wall is a rigid wall around the outside of the cell. It is made of tough fibers of cellulose that support the cell and give it its shape.

Chloroplasts are tiny structures that give plants their green color. They contain a green chemical called **chlorophyll** that plants use to help them make their own food.

The vacuole is a large space inside the cell that is filled with cell sap. This is a watery fluid containing dissolved **minerals** and food.

How many cells?

Some living things, such as human beings and trees, are made up of millions of cells. They are called multicellular, or many-celled, **organisms.** Other living things consist of only one cell. They are called unicellular, or single-celled, organisms. The amoeba is a tiny, unicellular organism that lives in water. Its cytoplasm flows along to make it move, and it feeds by engulfing its prey. To reproduce, an amoeba simply divides in two.

An amoeba is a unicellular organism.

Cell Specialists

Cells make up every part of a living thing, but not all cells look the same. You are made up of many different types of cells. For example, the cells that make your bones, nerves, or muscles have special features to help them to do a particular job. Cells do not work on their own, but are grouped together to build the different parts of your body.

Specialized cells

Cells come in different shapes and sizes, depending on the job they do. They are called specialized, which means that they can only do one type of work. Here are some examples of specialized cells:

- Nerve cells – Nerve cells, or **neurons,** have long, thin fibers for carrying messages all over your body.

- Red blood cells – Red blood cells carry **oxygen** around your body. They are doughnut shaped to give a large **surface area** so they can pick up as much oxygen as possible.

- Ciliated epithelial cells – These are flattened cells that form a thin layer to line your air passages. They are covered in tiny hairs called **cilia** that trap harmful dirt and germs.

Tiny cells called cilia line your air passages.

- Root hair cells – A plant's roots suck up vital water and **minerals** from the ground. The roots are covered in long, hair-like cells, like the ones in the picture, that provide a large surface area for taking in water.

6

Cells and systems

Groups of specialized cells make tissues, such as your muscle or bone tissue. Groups of different tissues make organs, like your heart or lungs. A group of organs working together is called a system, such as your digestive system. Digestion uses many organs, including your intestines and stomach. The systems work together to form an **organism,** like you!

Life processes

The cells in your body work together to keep you alive. They carry out the seven life processes, which are:

1. Movement – All living things can move their bodies.
2. **Respiration** – All living things use oxygen to release energy from food.
3. Sensitivity – All living things sense and respond to changes in the outside world.
4. Feeding – All living things need food for energy and growth.
5. Excretion – All living things must get rid of waste products from their bodies.
6. **Reproduction** – All living things produce young to replace those that die.
7. Growth – All living things grow. Most animals grow until they reach adult size. Some plants never stop growing.

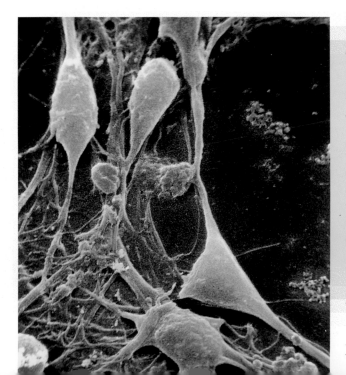

These brain cells will live longer than any of your other cells.

Did you know?

Some of the cells in your body last throughout your life and others only live for a few days. When they die, they are replaced with new cells. Some brain cells will last a lifetime, and your bone cells will last for about 30 years, but the cells that line your small intestine only live for two to three days.

Plant Systems

Plants range from tiny, single-celled **algae** to towering trees, made of many millions of cells. Most plant cells follow the basic pattern shown on page 5, but some are specialized for particular jobs, such as making food, taking in water, or carrying water around the plant. Groups of plant cells work together to form tissues, such as **xylem** and **phloem**. Groups of different tissues form organs, such as the plant's roots, stems, and leaves.

Parts of a plant

Flowering plants range from daisies and grasses to huge horse chestnut trees. These plants look very different from each other, but they all have a similar structure.

Leaves – Cells inside the leaves contain **chlorophyll,** which absorbs sunlight for **photosynthesis. Veins** in the leaves carry food and water around the plant.

Flowers – These contain the plant's sex cells for **reproduction.**

Stem – The stem supports the plant and holds the leaves up to the sunlight. It also carries food and water up and down the plant.

Roots – The roots anchor the plant in the ground and take in water and **minerals.**

Systems in a plant are all basically the same, whether it is a small flower or a huge tree.

Making food

All living things need energy to survive. This energy comes from their food. Green plants make their own food by **photosynthesis.** This takes place in the plant's leaf cells, inside tiny, disk-shaped structures called chloroplasts. The chloroplasts contain a green chemical called chlorophyll. It uses energy from sunlight to turn carbon dioxide from the air, and water from the ground, into a sugary food called **glucose.**

Respiration

To release energy from the glucose, plants use **oxygen** from the air. This is called **respiration,** and it happens inside the plant's cells. The plants take in oxygen through their leaves and give off carbon dioxide as waste. Respiration continues during the night when the plant cannot photosynthesize.

Flower functions

Flowers contain a plant's male parts, which produce **pollen** (the male sex cells), and female parts, which produce **ovules** (the female sex cells). For a new plant to grow, pollen must travel from the male to the female parts. This is called **pollination.** Then the **nuclei** of the pollen and ovule fuse, or join together, to make a new cell. This is called **fertilization.** The new cell grows into a seed that contains a new plant and a store of food.

Inside a leaf

Upper epidermis – This is the upper skin of the leaf. It is made up of a single layer of transparent cells that allow sunlight to pass through to the palisade cells.

Palisade cells – This is a layer of tall, column-like cells that are found just under the upper epidermis where they receive the most light.

Chloroplasts – These are found in the **cytoplasm** of the palisade cells. They contain chlorophyll for photosynthesis.

Lower epidermis – This is the lower skin of the leaf.

Stomata – These are tiny holes in the leaf for allowing water and gases in and out.

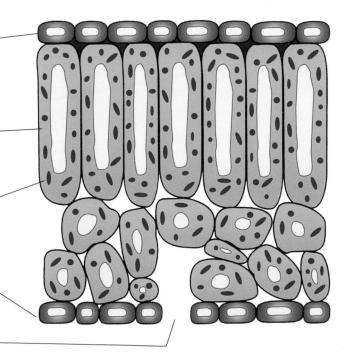

Plant Plumbing

Inside a plant, specialized cells work together to carry food and water around the plant in a constant flow. The water is sucked up from the ground through the plant's root system and carried up the stem to the plant's leaves.

Root hair cells

A plant's roots grow downward and sideways into the ground. Their tips are covered in thousands of tiny, tube-like structures, called root hairs. These are actually formed by the root's outer cells. They grow between the particles of soil, and absorb water and **minerals** by **osmosis.** Having so many root hairs greatly increases the roots' **surface area** so they can take in more water.

A plant's roots are covered in root hairs.

Transport system

From the roots, water is carried up the plant to the leaves through a system of tiny tubes called **xylem.** Another set of tubes, called **phloem,** carry food in the form of sticky **sap** from the leaves to the rest of the plant. All these tubes form the plant's "plumbing system." The xylem and phloem are arranged in bundles running through the plant's stem, like bundles of tiny drinking straws. Xylem and phloem are called **vascular tissue.** Some simple, non-flowering plants, such as mosses and **algae,** do not have vascular tissue.

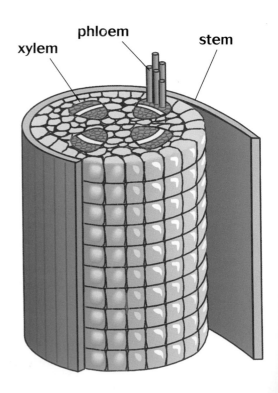

xylem phloem stem

Inside a Plant Stem

What is osmosis?

Water enters a plant's roots by osmosis. This is the way water moves from one cell to another. Some cells contain concentrated cell sap, with a large amount of **glucose** dissolved in a small amount of water. Others contain much weaker sap. In osmosis, water moves from the weak sap to the strong. Root hairs draw in water by osmosis because their cell sap is stronger than the water in the soil.

Did you know?

The wood in a tree trunk is made up of old xylem tubes. As the tree grows, the tubes of dead xylem cells in the center of the trunk fill with a hard, woody material called lignin. This supports the trunk and the tree's great weight. The younger, outer layer of xylem still carries water through the plant.

Losing water

Some of the water drawn up through the plant is used in **photosynthesis.** But most of it is lost through tiny holes on the underside of the leaves. The water then **evaporates** from the leaves into the air. The holes are called stomata. Each stoma, or hole, is surrounded by two oval-shaped guard cells, like the ones in the picture, which open and close the stoma.

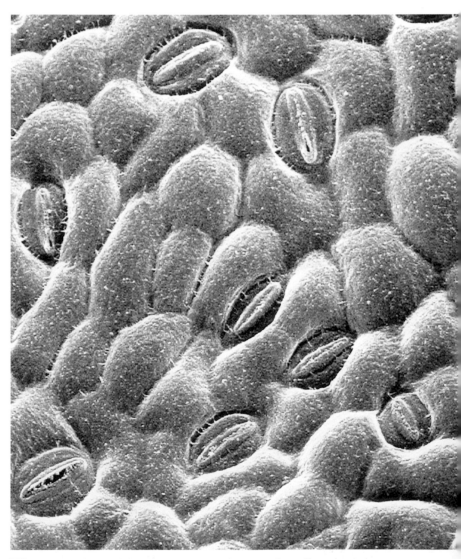

These stomata are on a rose leaf.

11

Bones

Like plants, all animals, including you, are made up of cells. Your body is like a complicated machine, with many different parts. Your cells, tissues, organs, and systems all work together to make you grow and keep you healthy, and to keep your body machine working. Most of the cells in your body are specialized, like the cells in plants. The cells that make the bones in your skeleton are specialized to help the bones do their job.

Skeleton support

Your skeleton is made up of a system of bones inside your body. It has three important jobs to do. First, it acts like the framework of beams and girders inside a building to hold your body up and give it its shape. Without a skeleton, your body would collapse. Second, your skeleton protects the organs of your body from being bumped and damaged. For example, the bones in your skull protect your delicate brain. Third, your bones anchor your muscles so that you can move.

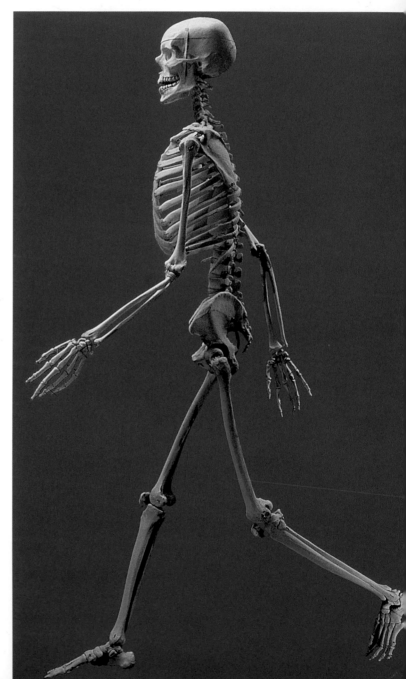

The human skeleton acts as support and protection for the body.

Tough bones

Because of all the jobs it has to do, bone tissue is highly specialized. The outer part of a bone is made of tough, non-living tissue, and is mostly **minerals,** such as **calcium.** This part of the bone is extremely strong to help the skeleton support your body weight and absorb bumps. The inside of the bone is living tissue that is soft and spongy. This makes bones light so that you can move, and flexible so that they do not snap or break too easily. In old age, people's bones have less living tissue so they become brittle and are easily broken. Some of the larger bones in your body contain a soft tissue called **marrow.** This makes your red blood cells.

The spongy tissue inside a bone is shown magnified here.

Growing bones

There are more than 200 bones in your skeleton. They come in a wide range of shapes and sizes, but they all have the same basic structure. When you were a baby, some of your cells formed tough, gristly **cartilage.** Gradually, the cartilage ossified, or turned into hard bone. This continues until you are about 25 years old. Some of the cartilage never turns to bone. You can still feel cartilage in the end of your nose and in your ears.

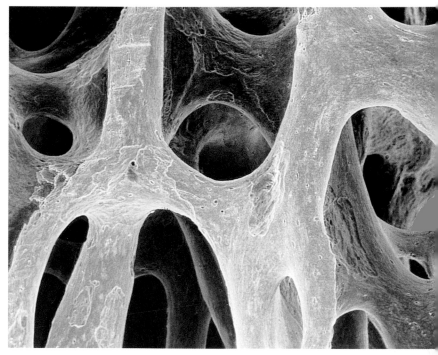

Did you know?

Your bones are covered in a thin, tough, outer layer of living bone cells called the periosteum. If you break a bone, cells in the periosteum divide and multiply, and grow over the break, joining the broken parts together. You might need a plaster cast to help the bones heal evenly.

Moving Muscles

There are hundreds of muscles all over your body—in your organs, such as your heart and bladder, or attached between your bones to allow you to move. Your muscles are a type of tissue. The muscles, shown in the picture, cover your skeleton. They are known as striated muscles because they look striped under a microscope. The muscles in your organs are called smooth muscles. Smooth muscles work all the time, automatically. A special smooth muscle, called **cardiac** muscle, keeps your heart beating.

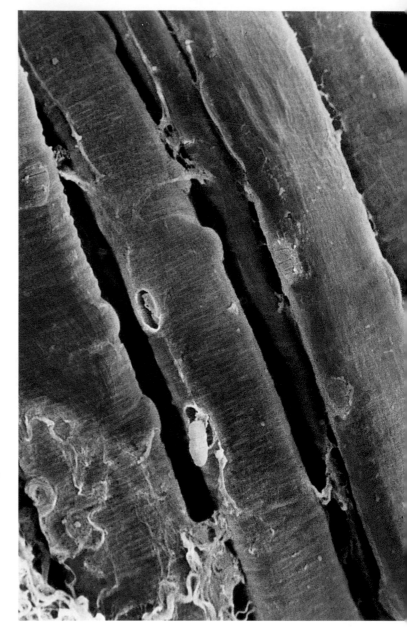

Muscle cells

Striated muscle tissue is made up of bundles of long, thin, thread-like cells, which are usually called muscle fibers. There are more than 2,000 fibers inside a large muscle. Each fiber is made up of even finer threads, called myofibrils. Muscle fibers can be up to 11 inches (30 centimeters) long. Their long, thin shape allows them to stretch so that you can move the different parts of your body. The whole muscle is covered in a protective, elastic skin called the epimysium.

14

How muscles work

Muscles are attached to your bones by strong straps called **tendons.** Muscles pull on your bones to make them move. When you want to move your arm your brain sends electrical signals to your muscles, telling the fibers to contract, or become shorter. This gives a pulling force. Muscles can only pull, not push, so most of your muscles work in pairs. In your upper arm, for example, the biceps muscle contracts to bend your elbow. Then it relaxes and its partner, the triceps, contracts to straighten your arm.

You can clearly see the muscles in this sprinter's legs.

Muscle power

Your muscles need a constant supply of **oxygen** and energy from food to keep them working properly. These are carried to your muscles in your blood, through tiny blood vessels that are wrapped around each bundle of muscle fibers. If supplies of oxygen and energy run low but your muscles are still working hard, they may go into a painful spasm. This is known as a cramp.

Did you know?

You have more than 600 muscles in your body. The largest are in your thighs and bottom, and are called the gluteus maximus muscles. Your smallest muscles are attached to the tiny bones deep inside your ears. They are called the stapedial muscles, and are about the size of a pinhead.

The Circulatory System

Your cells need energy from food and **oxygen** from the air to keep them working properly. These are carried around your body by your blood, through blood vessels. Your blood also collects waste products for disposal. Your blood is pumped by your heart, and circulates continuously around your body. Your heart, blood, and blood vessels make up your circulatory system.

Blood cells

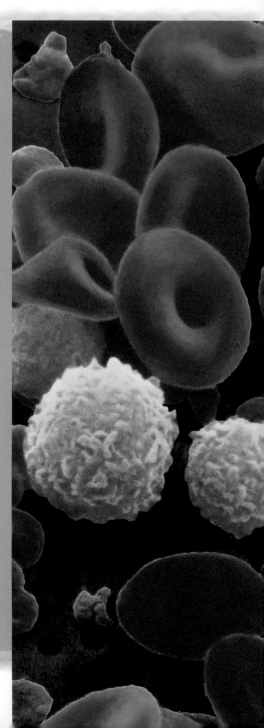

Blood is made up of a clear, straw-colored liquid called plasma. Red blood cells, white blood cells, and platelets all float in your blood, as you can see in the picture. You have a little more than one gallon (5 liters) of blood in your body.

- Plasma – Plasma is made up of water in which **proteins,** salts, food, and waste materials are dissolved.
- Red blood cells – Red blood cells carry oxygen from your lungs. They are made in the soft **marrow** inside your large bones. Red blood cells contain a chemical called **hemoglobin.** This is what makes blood red. It absorbs oxygen in your lungs, then your blood flows to your heart to be pumped around your body.
- White blood cells – White blood cells help your body fight disease. Some eat up harmful germs, which enter your body through cuts, food, or air. Others make chemicals called **antibodies.** The antibodies stick to germs and kill them.
- Platelets – Platelets are tiny fragments of cells that have broken off from larger cells in the bone marrow. They help your blood to clot so that you do not lose too much of it when you cut yourself.

Blood vessels

Blood travels in tubes called blood vessels. You have so many of these that, put end to end, they would stretch more than twice around the earth. **Arteries** are strong, muscular tubes that carry blood from your heart. They divide into tiny **capillaries** that reach your body cells. The walls of the capillaries are only one cell thick, so substances can easily pass between them and your cells. The capillaries join up again to form **veins,** which return the blood to your heart.

The Circulatory System of the Human Body

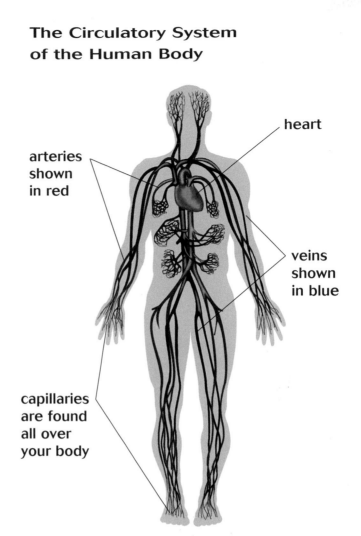

heart

arteries shown in red

veins shown in blue

capillaries are found all over your body

The heart

Your heart is an organ. It works like a pump, pushing blood around your body. It pumps, or beats, about once every second and, with each beat, sends blood surging along the arteries. Your heart is made of strong **cardiac** muscle. As the muscle contracts, it squeezes blood out and around your body. Each contraction is called a heartbeat. During exercise, your heart has to beat faster to supply your muscles with extra energy. Your **pulse** rate measures how often your heart beats per minute.

Did you know?
A pinprick of blood contains an amazing 2,500,000 red blood cells, 5,000 white blood cells, and 250,000 platelets. In total, you have about 30 billion red blood cells in your body, more than any other type of cell.

The Digestive System

Everything you do uses energy. Walking and running use lots of energy. But breathing, blinking, and even standing still use energy too. You get this energy from the food you eat. Food also supplies all the chemicals you need for growth and to keep your body working properly. The useful parts of food are carried to your cells by your blood. But only tiny **molecules** of food can pass into your blood, so first, the food must be broken down. As the food passes through your body it is broken down by a certain set of organs. This process is called **digestion.**

Digesting a meal

1. Your teeth and tongue chew and mash your food. It is mixed with **saliva** so it is easier to swallow.
2. The food goes down your **esophagus** into your stomach. It is pushed along by the smooth muscles of the esophagus walls.
3. In your stomach, the food is mixed with digestive juices that contain special **enzymes.** It forms a creamy mixture.
4. The food passes into the first part of your small intestine. A green liquid, called bile, is made in your liver. It breaks down any fat into tiny droplets. Other enzymes, made in your **pancreas,** break the food down even more.
5. In the second part of the small intestine, called the ileum, the digested food is absorbed into your blood.
6. Any undigested food goes into your large intestine and is passed out of your body as solid waste called **feces.**

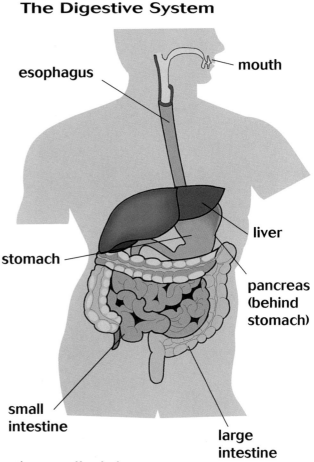

The Digestive System

esophagus

mouth

liver

stomach

pancreas (behind stomach)

small intestine

large intestine

How food gets into your blood

The inner wall of the ileum is covered in tiny, finger-like structures called villi. These make a huge **surface area** for absorbing food. The villi walls are only one cell thick, so food can pass through them and into the tiny blood vessels that carry it around your body to your cells.

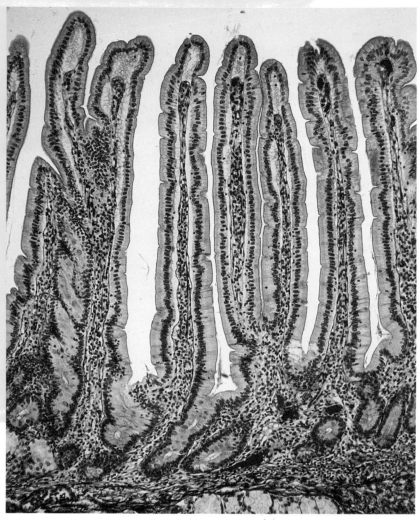

These magnified villi are clearly only one cell thick.

Speedy enzymes

Enzymes are special chemicals made in your cells. They are called catalysts, which means that they can speed up the chemical reactions taking place in your body. There are thousands of different types of enzymes. Digestive enzymes are only one type, which help to break down and dissolve your food as it passes through your digestive system.

Did you know?

Your digestive system is about 30 feet (9 meters) long, and a meal takes about three days to pass through it. Your small intestine alone is about 20 feet (6 meters) long. It is called small because it is skinnier than the large intestine.

The Respiratory System

Your body needs **oxygen** to release energy from your food. When you breathe in, you take air into your lungs. Here the oxygen is removed, and carried to all the cells in your body by your blood. Inside the cells, the oxygen is used to release energy from food in order to keep your cells alive and working. This process results in your cells putting out waste made of **carbon dioxide** and water, which leaves your body when you breathe out. This whole process is called **respiration.** It is easy to confuse respiration with breathing, but breathing is only part of respiration. It describes only the first action, taking air into your lungs, and the last action, breathing carbon dioxide out of them.

The bronchioles branch off of the lung's bronchial tubes.

Breathing in

When you breathe in, air is sucked in through your nose or mouth. It goes down a large tube called your **trachea** and into two tubes, called bronchial tubes, that lead into your lungs. The bronchial tubes gradually divide to form a network of tiny tubes, called the bronchioles. As you breathe in, your chest muscles move up and out to let your chest expand so that your lungs can fill with air.

The Respiratory System

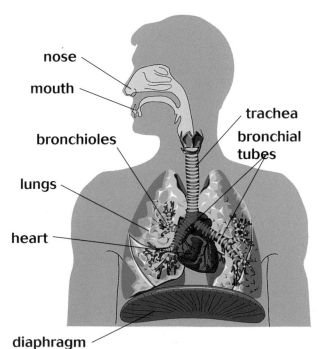

nose
mouth
trachea
bronchioles
bronchial tubes
lungs
heart
diaphragm

Gas exchange

At the end of the smallest branches of the bronchioles are bunches of tiny air pockets called alveoli. When you breathe in, these fill with air like tiny balloons. The alveoli are covered with very fine blood vessels. In places, their walls are only one cell thick. Oxygen from the air passes through the walls and into your blood. Then it is carried around your body to your cells, where it is used to release energy. The millions of alveoli in each lung form a huge **surface area** that absorbs the oxygen you inhale.

Breathing out

Your blood also carries carbon dioxide waste from your cells to your lungs to be breathed out. The carbon dioxide leaves the blood and moves into your lungs through the alvioli. It is drawn up through your bronchial tubes and trachea, and out through your nose or mouth. When you breathe out, your chest muscles relax. This lowers your chest and raises your **diaphragm** to force the air out of your lungs.

The alveoli inside a lung are like tiny balloons.

Did you know?

Two special types of cells line your air passages. One type is covered in tiny hairs called **cilia.** The other produces slimy **mucus.** Dust and germs stick to the mucus, then the cilia beat them to the back of your mouth so that you can swallow and clear the germs away. People who smoke often suffer from painful breathing diseases like **bronchitis.** This is because the chemicals in cigarette smoke prevent the cilia from doing their job properly.

Water and Waste

All the chemical processes happening in your cells make waste products, such as **carbon dioxide,** undigested food, and water. All of these must be removed from your body so that they do not harm your cells. You breathe out carbon dioxide during **respiration,** and get rid of waste food as **feces.** You lose waste water when you sweat or **urinate.** Your body is about two-thirds water, and this balance must be kept for your cells to function properly. If you take in more water than you need, your kidneys make **urine** to get rid of the excess liquid and other waste products.

The urinary system

Your kidneys and bladder are known as your urinary system because they make and store urine. Your two kidneys are in your lower back, level with your waist. As blood passes through them, the kidneys filter out waste water and other substances. This waste liquid is called urine. It flows down two tubes, called ureters, into your bladder, an organ that is like a stretchy, muscular bag. The urine is stored in your bladder until you go to the bathroom. Then it passes out of your body through a long tube called the urethra.

The Urinary System

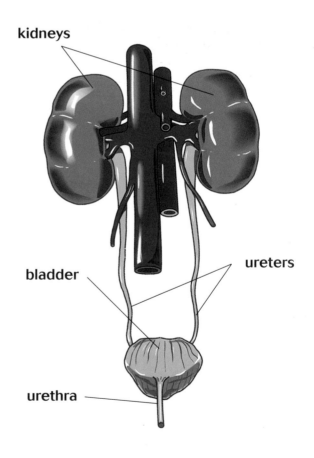

kidneys

ureters

bladder

urethra

Filter system

Each kidney contains over a million tiny filtering units called nephrons. Blood flows into your kidneys through your **renal artery.** When it has been cleaned, the blood flows out through your **renal vein**, returns to your heart, and continues to circulate around your body.

Nephrons are located at the ends of the smallest arteries in the kidney.

Poison control

Your liver is the largest organ in your body, weighing over 3 pounds (1.5 kilograms). It does several very important jobs. When your blood has absorbed digested food, it makes a detour through your liver. Your liver stores some of the **nutrients** in the food and changes others into more useful forms. It also gets rid of the toxic, or poisonous, substances found in some food and drink, by turning them into harmless substances. This process is called detoxification.

Your Senses

Your five senses are sight, hearing, smell, taste, and touch. They tell you about the outside world. You receive information through special **sensory** nerve cells called **receptors.** They react to changes in your surroundings like light or temperature, and send messages to your brain. There, the information is processed and your brain tells you what is happening. Many receptors are grouped to form sense organs, such as your eyes, ears, or nose.

Eyes and seeing

You see things because light from an object enters your eyes. It is focused by the cornea and lens, which you can see in the diagram, to project a sharp, upside-down image onto the retina. Millions of light-sensitive cells on the retina react to the light and send messages to your brain, which interprets the signals and produces the picture you see.

A Cross-section of a Human Eye

iris – the colored part of the eye

pupil – light enters here

lens – focuses the image

cornea – helps focus the image and protect the eye

muscles – move the eye

retina – area with nerve cells

optic nerve – main nerve to brain

Did you know?

The light-sensitive receptor cells in the retinas of your eyes are called rods and cones. Rods detect black and white, and how bright the light is. Cones detect colors. Some people are color blind. They cannot see the difference between certain colors, because their cone cells are not working correctly.

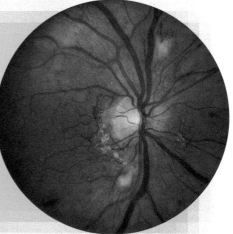

This is a human retina.

Ears and hearing

Sounds are vibrations, or waves, in the air. Your outer ears funnel them down your ear canal to the eardrum. This is a thin **membrane** that vibrates as the sound waves hit it. The vibrations are passed on to three tiny bones, then onto another membrane, and then to the **cochlea**. The cochlea is filled with liquid that also vibrates. Special nerve cells detect the vibrations and turn them into signals, which travel to your brain. Here they are turned into the sounds that you hear.

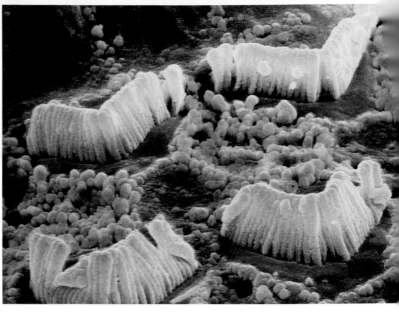

These cells are specialized to sense sound.

Taste and smell

Your tongue and nose are your organs of taste and smell. Your tongue is covered in taste buds, visible in the picture, which are lined with taste-sensitive cells. These cells send messages about taste to the nerves. Smells are chemicals floating in the air. They travel up your nose and are picked up by smell receptor cells. Taste and smell are closely linked. If you have a cold and your nose is blocked, you may find it difficult to taste your food.

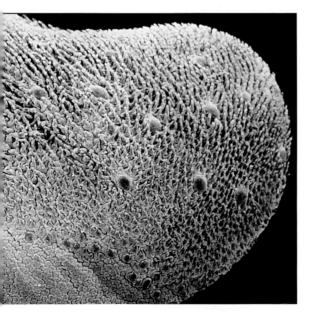

Skin and touch

Millions of sense receptors are packed beneath the surface of your skin and each type detects a different sensation. Some are sensitive to touch, temperature, and pressure. Some tell you if the things you feel are rough, smooth, hard, or soft. Other receptors detect pain, which warns you that something is wrong with your body.

The Nervous System

Your brain, nerves, and **spinal cord** form your body's central nervous system. This system processes the information your senses receive about the outside world, and instructs your body how to think, feel, or react. Your nerves are like long, fine wires that carry messages in the form of electrical signals between your brain and the other parts of your body. Your spinal cord acts as the main pathway for these messages to travel on as they move from the nerves, to the brain, and then back to the body.

Your amazing brain

Your brain controls every part of your body, and everything that you think, learn, remember, and feel. It has up to 10 billion nerve cells, or **neurons**. Nerves carry information about the outside world from your senses to your brain, where it is sorted and processed. Then your brain sends messages along your nerves to your body to tell you what action, if any, to take.

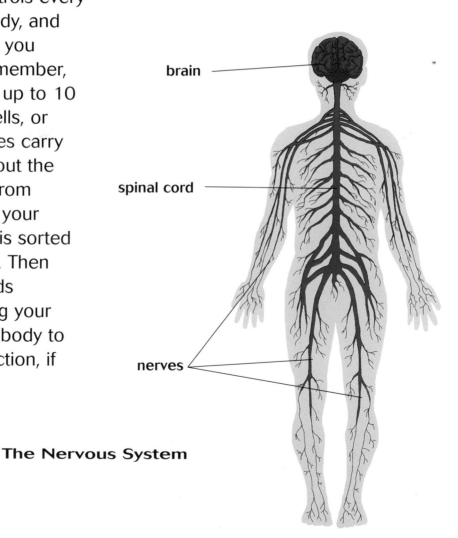

brain

spinal cord

nerves

The Nervous System

Nerve cells

A nerve is made up of bunches of long fibers, called axons, that run from the cell bodies of the neurons. Shorter fibers, called dendrites, also branch off from the cell bodies. You have a vast network of about 100 billion neurons running throughout your body. There are two main types of neuron. **Sensory** neurons carry messages from your sensory organs to your central nervous system. **Motor** neurons carry messages from your central nervous system to your muscles to make them move.

How nerves work

One neuron's axons lie very close to the next neuron's dendrites, but they do not actually touch. Nerve signals pass from one neuron to another by jumping over a tiny gap called a synapse, shown in the picture. To cross the gap, the signals must change from electrical signals into chemicals called neurotransmitters. Once the signals have crossed, they turn back into electrical signals again. One neuron has synapses with many other neurons. This means that it can receive many different electrical signals.

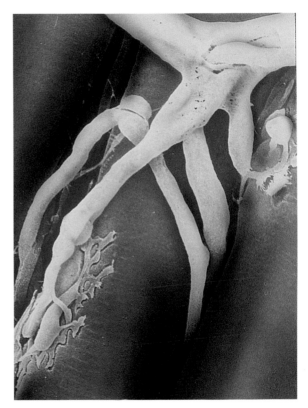

Did you know?
Some of the neurons in your body will last a lifetime, but not all of them. Thousands of your brain and nerve cells die every day. In fact, you start losing them even before you are born. Unlike many other types of cells, neurons can never be repaired or replaced. Fortunately, you have millions more, so you will never run out!

Reproduction

Reproduction means the creation of new life. Human beings produce sexually, which means that there are two parents who each make sex cells. Male sex cells are called sperm. Female sex cells are called ova, or eggs. For a baby to develop, a sperm must join with an ovum to form a new cell. This is called **fertilization.**

The first cell division of a fertilized egg the first stage of the baby's growth.

Reproductive systems

Your reproductive system starts working at **puberty.** This usually begins at age 11 to 13 for girls, and age 12 to 14 for boys. An ovum is activated in one of a female's two **ovaries.** It then travels down a tube, called the Fallopian tube. If it meets a sperm cell, it may be fertilized. Sperm cells are made in a male's **testes.** They swim along two tiny tubes, called sperm ducts, to his penis.

Male and female

A baby develops when a sperm meets and joins with an ovum in the female's body. This happens when a male and female have sexual intercourse. After fertilization, the two cells make one new cell, which begins to divide until it forms a ball of cells. This embeds itself in the female's uterus, or womb. Over the next nine months, this ball of cells grows and develops into a baby.

28

Cell instructions

Each sex cell carries instructions called genes. Genes are information like codes that say what a living thing is and what it looks like. They determine the characteristics you inherit from your parents, such as hair and eye color. They are carried on very fine threads of material, called chromosomes, inside the **nuclei** of your cells. You inherit 23 chromosomes from each parent. The chromosomes that determine your sex are X and Y chromosomes. All ova carry X chromosomes. Half the sperm carry X chromosomes and half carry Y. If two X chromosomes join, the baby will be a girl. If an X and a Y chromosome join, the baby will be a boy. The chromosomes shown in the picture, or karyotype, are a male's. You can see the XY pairing in the bottom right corner of the karyotype.

Cell division

For a living thing to grow bigger, new cells must be made. Most cells reproduce by splitting in two. There are two types of cell division:

- Mitosis – The nucleus splits in two, forming two new daughter cells. Each of these cells has a full set of chromosomes. The two cells are identical. This type of division is used for growth and repair.

- Meiosis – The nucleus splits in two, forming two new daughter cells. But each of these cells has only half the total number of chromosomes of its parent cell. The two cells are not identical. This type of division is used to produce sex cells for reproduction.

Conclusion

Cells are the building blocks of life. Your own body is made of millions of cells. Without cells and the systems that they form, you would not be able to move, eat and digest food, or read this book. Your amazing cells make your body, and make it function.

Glossary

algae   one-celled plant found in both saltwater and freshwater

antibody   chemical, made by some types of white blood cells, that sticks onto germs and kills them

artery   blood vessel that carries blood away from your heart

bronchitis   chest infection that makes breathing difficult and painful

calcium   mineral that is essential for building your bones

capillary   smallest blood vessel

carbon dioxide   gas that living things breathe out as waste during respiration—plants use carbon dioxide in photosynthesis

cardiac   to do with the heart

cartilage   rubbery, flexible tissue

cellulose   tough material made of fibers found in plant cell walls

chlorophyll   green coloring found inside plant cells that absorbs energy from sunlight for use in photosynthesis

cilia   tiny hair-like structures that cover some cells

cochlea   curled tube, filled with liquid, in the inner ear—sound waves shake the liquid, which pulls on nerve endings to send signals to the brain

cytoplasm   gelatin-like substance that fills cells

diaphragm   muscle under your chest that contracts and relaxes so you can breathe in and out

digestion   way in which food is broken down and processed as it passes through your body

enzyme   chemical made in your cells—some enzymes help break down your food during digestion

esophagus   large tube in your throat through which you swallow food, also called the gullet

evaporate   when a liquid, like water, turns into gas

feces   solid waste you pass when you go to the bathroom

fertilization   joining together of a male and female sex cell to produce a new living thing

gelatin   squishy food like jello

glucose   simple sugar—organisms store food as glucose

hemoglobin   chemical in your blood that makes it look red and carries oxygen around your body

marrow   soft substance inside some of your larger bones that makes new red and white blood cells

membrane   thin sheet of tissue

mineral   substance that helps to build your body and keep it healthy

molecule   tiny particles of a substance

motor   motor neurons are nerve cells that carry messages from your brain to your muscles

mucus   slimy substance that helps protect the lining of your nose, lungs, and stomach

neuron   nerve cell

nucleus   rounded structure inside a cell that is the cell's control center— nuclei is the plural of nucleus

nutrient substance in food that your body needs to function

organism scientific word for a living thing

osmosis way in which fluid and chemicals pass from one cell to another

ovary part of a female's reproductive system—the two ovaries make egg cells

ovule female sex cell of a plant—after fertilization, they become seeds

oxygen gas that all living things need to take in to survive

pancreas group of cells that is part of your digestive system—the pancreas makes enzymes that help break down your food

phloem tubes that carry food through a plant

photosynthesis process by which green plants make their own food from carbon dioxide and water, using energy from sunlight absorbed by their chlorophyll

pollen tiny grains that are the male sex cells of plants

pollination transfer of pollen from a male flower to a female flower, or from male to female parts of a flower

protein chemical substance that living things use for growth

puberty time in a girl or boy's life when their bodies change from being a child to being an adult

pulse the beats of your heart—you can feel a pulse in your wrist or neck

receptor special nerve cell that senses the outside world and sends messages about it to your brain

renal to do with the kidneys

reproduction creation of new life

respiration way you breathe in oxygen, used to release energy from your food, and breathe out carbon dioxide as waste

saliva liquid made in your mouth that helps to break down your food

sap liquid inside a plant that carries food and water

sensory sensory neurons are nerve cells that carry messages from your sense organs to your brain

spinal cord thick bundle of nerves running down your back, inside your spine or backbone

surface area space taken up by the outside of something

tendons thick straps of tissue that connect muscles to bones

testes part of a male's reproductive system where sperm cells are made

trachea long tube running down your throat used in breathing, also called your windpipe

urinate to pass liquid waste when you go to the bathroom

urine liquid you pass out of your body that contains waste water and other waste substances

vacuole space in the middle of a plant cell

vascular tissue system that carries sap around the plant, made of xylem and phloem

veins **1)** tiny tubes in a leaf that carry food and water **2)** vessels that carry blood back to your heart

xylem tubes that carry water through a plant

More Books to Read

Balkwill, Fran. *Cells Are Us*. Minneapolis, Minn.: Lerner Publishing Group, 1996.

Nye, Bill. *Bill Nye the Science Guy's Human Body Book*. New York: Hyperion, 2000.

Roca, Muria Bosch and Marta Serrano. *Cells, Genes & Chromosomes*. Broomall, Pa.: Chelsea House Publishers, 1995.

Index